THE
INTELLIGENCE OF
DUMB ANIMALS

THE INTELLIGENCE OF

DUMB ANIMALS

Ranching experiences document that animals are very intelligent.

Ken Bull

To order additional copies of this book, contact:
Xlibris Corporation
1-888-795-4274
www.Xlibris.com
Orders@Xlibris.com
67615

CONTENTS

Foreword.. 7

Acknowledgments 9

Prologue: The Intelligence of Dumb Animals............................ 11

      What is Ranching? 13

Introduction... 17

Chapter 1: Horses 19

Chapter 2: Cows 29

      Birthing Instincts 31

      Twin Calves.. 33

      Parental Herd Instincts... 35

      One Fine Aggie Cow 36

      Telepathic.. 37

      Agriculture Education 38

Chapter 3: Sheep... 39

Chapter 4: Sheep Dogs... 43

Chapter 5: Goats... 55

      Nancy Goat.. 58

      Cabrito, Ole 59

Chapter 6: Wildlife 61

      Raccoons.. 64

      Mean Coon 64

      Cats... 66

      Black Cats 71

      Blind Cat 72

      Deer.. 75

      Coyotes 78

      Turkey... 80

Conclusion... 83

About the Author 85

FOREWORD

We often refer to nonhuman vertebrates as dumb animals. We ranchers, who work closely with animals, know that animals are not dumb.

Ranching is a business that provides food for Americans by properly fitting animals to the land so that the plants from nature are converted to food for humans. There are all sizes of ranches, some are large and some are small. The small, Mom and Pop, ranches provide most of the food but receives little of the credit.

Over the years on our small ranch, we have raised cows, sheep and goats out here amongst the deer, turkey, coyotes, raccoons, squirrels, opossums, armadillos and rattle snakes.

To manage the producing animals, we solicit the help of other intelligent animals. All ranches must have horses to corral the livestock. Horses also serve as entertainment for the visiting grand kids. Dogs of certain species that are properly trained help protect the livestock from predators. Probably the hardest working support animals are the cats that habitat the barn area to control the mice and rat populations that lure in rattle snakes. Cats also serve as garbage disposals since they will eat all of the leftovers from our meals.

Barbara and I have spent many years operating her small ranch in Central Texas. The ranch is located 12 miles north of Brady, Texas. The stories I document here are our actual experiences. We ranch not so much for profit but because we love the animals and our direct contact with them has taught us that animals are not dumb.

ACKNOWLEDGMENTS

I want to make it clear that while this book shows me as author the experiences are both mine and Barbara's. She owns the ranch and all decisions that we have made were made by joint considerations.

Barbara grew up here on this ranch although they lived in Brady. Her dad took her with him to help on the ranch long before she was old enough to really help. She learned a lot about ranching long before I had the privilege of being her husband.

After we were married, we spent a career in the United States Air Force before coming back here to operate this ranch. She also has the experience of raising two wonderful children.

I want to also acknowledge our children. Our daughter Karen not only helped us during part of her life but went on to mother two beautiful daughters. Our son Kenneth was also a help during his early years and now is the father of two excellent sons.

I also thank God for creating this planet of animals and allowing me to be a part of this amazing experience. Thank You God for helping me record these experiences.

PROLOGUE

The Intelligence of Dumb Animals

Are animals really dumb? I guess it depends on what the definition of dumb is. If dumb means that animals do not make noises, then they are not dumb. If dumb means that animals do not communicate, then they are not dumb. If dumb means that they do not communicate with humans, then there is some doubt, but one might question whether the problem is with the animal or with the human. Certainly they communicate with others of their species and often with other species as well. I think that they can and do communicate with humans.

Most of those who read this will know that their pets do communicate with them. A dog or a cat will surely get across that they need to "go out." Dog owners know that dogs learn tricks and perform on command. Cat owners know that cats spend all of their time training their owners. Certainly pets understand human languages. They know when you are unhappy and when you are pleased. Your gruff voice may send them cowering but they know that their careful response to your anger usually will bring forgiveness.

The lesser domesticated animals, horses, cows, sheep, goats, dogs, cats and the wild animals as well do respond to humans. They respond to voices and are particularly attentive to body language.

We are told that if we were to encounter a bear that we should not retreat or run because the bear will then attack. We are supposed to charge the bear with raised arms and use a loud voice to frighten it away. I surely hope that I never have to face that decision.

Barbara and I own and operate a ranch in Central Texas. Actually, Barbara owns the ranch and I just work here. Truthfully, the ranch runs us. It decides what we will do each day and when we might take a trip or even when we may go to town for more supplies. Invariably, when we leave for a day or two, the bulls will break down a gate and let the herds mix or the horses will get into a wire fence and cut a leg or some cow will get in trouble having a calf. One gets afraid to leave the place. I don't know how the animals know that we are gone but they seem to know.

But it's a great life style if you love working with animals. We took over the ranch operations in 1974 and now live in a house on the ranch. We have raised horses, cows, sheep, goats, dogs and cats out here amongst the deer, racoons, armadillos, opossums, bobcats, coyotes, snakes and mountain lions. Some years we make a little money depending upon the livestock market, the weather and the government. More years than we would like, we end up with little income and chalk it off as another year of great experience and forced intense exercise.

Over the years of living and working with animals we have noted that animals are intelligent and far from being dumb. We notice many human traits in animals or perhaps animal traits in humans. They live by a hierarchy. There is a boss cow or boss horse or a boss deer and the others within that herd yield to that hierarchy. There is also a dominance of one animal species over another, including humans. The animals here on the ranch have become accustomed to our way of handling them and are accustomed to our appearance. They know how we usually dress, jeans and cowboy hats, and respond differently if we are not dressed in the customary garb. They definitely notice strangers and will be far more difficult to handle with strangers around. Some are quick to gentle while others are inherently wild. Some are funny to watch while others are always dull. They definitely grieve over the loss of an offspring. Unlike humans, the newly born animals instinctively know to get up, nurse and make their own way.

The stories that I document here are actual experiences that Barbara and I have had here on the ranch. We love the animals and believe that we are not alone. Most mom and pop ranching families are in business because they love animals and love the ranch way of life. We know that they do not stay in the ranching business because of profit. The ever-increasing cost of operations has long ago taken the profit from ranching. America owes a lot to small agriculture operations because these small ag businesses have kept the cost of food and fiber reasonable for the American families. Contrary to the belief of some folks in the cities, food does not grow in the grocery store. The grocery does have a lot to say about the price the folks pay but the food comes from the farms and ranches.

We hear a lot today about animal rights and how the cruel ranchers are endangering animal species. I hope that these stories show that ranchers are by no means cruel to animals. Animals are greatly respected and appreciated by the small family ranchers. We very much respect the intelligence of our dumb animals.

Please excuse the occasional use of cowboy poetry. I have found that poetry is an excellent way to express a view without having to worry with the proper use of English grammar.

What is Ranching?

Sitting here in my favorite chair
Looking out at the oak trees
Relaxing without a care
The limbs swaying in the breeze.
Near the house a yard serene
Then as far as the eye can see
Grass growing beautifully green
Beneath each and every tree.
That's Ranching.

There are no city lights
The traffic does not howl
Stars are beautiful at night
You can hear the call of the owl.
Neighboring squirrels are seen at play
Wild turkey come around
White tail deer seem to say
Put more corn on the ground.
That's Ranching.

Then there are times we all fear
When the rains do not come
The sky is far too clear
Earth baked from too much sun.
The weeds will have their way
Good grasses cannot grow
Fields are bare, there's no hay
But rain will come that we know.
That's Ranching.

No matter the conditions or mood
The rancher must heed a plan
Provide the fiber and the food
For each and every American.
So management is the key
Though not easy to understand
To ranch successfully
Fit the animals to the land.
That's Ranching.

INTRODUCTION

Ranching is a business of marketing the vegetation grown on the ranch whether it is the grass, weeds or browse. Cattle, for example, are a means of marketing the grasses grown on the ranch land. The cows eat the grasses and produce milk. Their calves nurse the milk, grow up and are sold for beef. Sheep convert the grasses and weeds to wool that is then marketed.

Also, the meat of lamb is used for food in some parts of the world.

Goats eat some grass and weeds but are mainly raised to market the browse from the tender new growth of brush and trees.

The young, tender, goat kids make excellent barbecue (cabrito). In addition, goats help manage the pastures by keeping the brush under control thus allowing for a balance of grass, weeds and browse.

Barbara's dad told us that a successful rancher raises sheep and goats to pay the lease and cows for the fun of it. We've found that he was right. Ranch land will produce more profitably by marketing the plants through a mixture of animals. Cattle, sheep and goats are compatible in that they consume a different plant and do not directly compete for the same plants. The land will support more "animal units" with a mixture of animals than with one species alone. The trouble is that sheep and goats are small animals and are the favorite food of several predators. The coyote and bobcat love lamb and kid goats. The animal rightist groups have brought about government regulations that make it difficult to control predation. As a result, the ranchers have problems with predators that make it unprofitable to raise sheep and goats.

The grandest of all is the horse. The horse does eat grass but it does much more. The horse serves as a management tool so the rancher can control the other animals that are converting the grass to sellable produce

CHAPTER 1

Horses

Every ranch has to have at least one horse. You can't "qualify" as a ranch in Texas without a horse and it doesn't matter whether or not the ranch needs a horse.

At least that's what Barb told me. I should have taken into account that she is a horse lover and has had horses all of her life. She is an avid reader of Blood Horse magazines and knows equine bloodlines like a trainer. Years ago, we visited the horse farms in Kentucky and the grooms loved her because they could talk horses with her——not like most of us tourists. She would ask about a particular horse she knew was standing as a stud at that farm and the groom would say "wait here." Soon he would lead out the horse she had asked about and show it to her. That made me really proud of her and her knowledge of horses.

Even now we often visit Lone Star Park when she knows that good horses are racing. She never places a bet——just enjoys the horses. She especially likes the turf races with the well-groomed racehorses performing on the beautiful green grass. I like the beautiful horses but I like the one most that wins while honoring my small wager.

Barb grew up here on the ranch that we now operate. She was the older of two girls and she served as the "boy" her dad wanted but did not produce. He took her with him to help with the ranch work long before she was old enough to really help. He would keep a horse for her to ride, some of them were gentle and some were not. But to her they were all intelligent.

She tells of a time when a smart horse also got her in trouble. She had been watching too many western movies and wanted to follow in the steps of some of the Hollywood cowboys. She saddled her horse to ride while her dad worked around the pens. She rode to a nearby water pond and decided to ride her horse across as they often did in the movies. She removed the saddle so not to get it wet and make her dad angry. The idea was to enjoy the experience, re-saddle and get back to the pens and her dad would never know. The horse was not trained to cross in water and began to lunge in the mud. She tried to hang onto his mane but soon was thrown off. She had lots' of experience swimming in muddy ponds and had no trouble getting out on the bank in very wet clothes. In the meantime, the horse quickly left the pond and ran back to the barn to report the incident to Barb's dad. Her dad, of course, seeing a wet horse without a saddle and no Barb, became very concerned and started searching the ranch ponds. He quickly found Barb starting to walk toward the pens. He listened and showed concern, but deep down he was sort of proud.

Here on the ranch, every horse must have a name. We've had Punkin, Gypsy, Blondie, Lady, Lil, Smokey, Socks, Bill and others beyond my memory. Later we had Granny and Fancy, a couple of very nice mares. We didn't need either of them since we were no longer raising sheep or goats.

A horse is not needed for cattle unless you are a young macho cowboy and those days are long past for us. Cows can be taught to follow the pickup to the pens so long as you feed them once they are in the pen. The rattle of cow-cubes in a paper sack is just like the ringing of the dinner bell to a hungry (spoiled) cow. Occasionally a horse is needed when the calves are young and the cow doesn't want to take the new calf to the pen or when a wayward bull has gone next door to visit the neighbor's cows and you have to go bring him home.

I must admit that horses may be needed for grand kids. A red-blooded rancher grandpa may want to impress the grand kids by letting them ride when they visit. Of course, the grand kids love the experience.

Horses are by far the most intelligent animals on the ranch. They can be taught to do tricks or help with the ranch work. All it takes is a rancher who knows more than the horse and one who uses that extra food-treat in just the right way. The crack of a whip or touch of a spur has been known to aid the learning process or refresh the memory.

As with Barb's pond swimming incident, it's the things the horses come up with on their own that shows their true intelligence. I remember one such event back in the Punkin and Gypsy days. Normal practice is for us to keep the horses in the horse trap, a 140-acre fenced pasture that abuts the pens. For some reason, we had moved them into the 540-acre East Pasture where the house and barn are located. Normally they would be around the barn each morning waiting for their morning ration. Any morning that they did not come to the barn always raised some degree of alarm, but soon they would show and everything would be all right.

One morning, before we left the house for the pens, Punkin came to the house nickering and looking back toward the east. We immediately knew that something was wrong with Gypsy or they would have been together. As soon as we started for the pickup to begin the search, Punkin headed east at a long trot. We followed like trained owners should follow and sure enough she led us to Gypsy.

Gypsy had been visiting neighbor horses across the fence and had gotten a front leg hung in such a way that each time she tried to free herself she would cut her leg on the barbed wire. Her leg was high up in the wire so that she could not place her foot on the ground. She was bleeding badly and was very tired from standing on just three legs.

We knew that a veterinarian would be needed. We cut her free from the wire and tied on a temporary bandage to control the bleeding. Barb stayed with her for comfort, both her's and Gypsy's, while I went to get a trailer. Gypsy gladly loaded in the trailer knowing that we were going to help her. After several stitches provided by a veterinarian and lots of love and extra feed provided by us, she was ready to go visit the neighbor horses again. This time she learned to visit from a distance. Punkin got a little extra feed from us and a gentle thank you from Gypsy for being smart enough to go for help.

Horses show compassion for their horse friends and their human friends as well. Our daughter, Karen, owned a black mare that she called Lady. Lady showed compassion for humans on more than one occasion.

We had "interviewed" several horses when we were searching for Karen's horse. There may be more "kid horses" out there than we think but it certainly pays to shop thoroughly before buying one. Horses can show intelligence in more ways than just kindness to humans and will often take advantage of inexperienced riders. They can definitely detect experience or lack thereof and sometimes react with a bit of mischief.

One horse we interviewed turned out not to be the "kid horse" that the owners advertised. When we went to see the horse we noticed that she was overly fat and the owner had her quite sweaty by the time we arrived. That aroused our suspicion. As I got on the horse to take her for our first ride, I noticed that the owner held onto the bridle while I got in the saddle. I rode off a short distance and decided that I would dismount and remount to see why he had held her head for me. I found out in a hurry. About the time my right leg crossed over the saddle while remounting, she started bucking and threw me off. I have a bent middle finger on my right hand to this day to remind me of that event. Fortunately, I managed to hold on to the rein and once cautious managed to get on successfully and ride her back to the owner. Barb had watched the rodeo and had already told the owner before I got back that she didn't believe that was the "kid horse" we were looking for.

Providence did lead us to Lady, a very nice looking black mare, that turned out to be a super "kid horse" for Karen. Lady was a well-bred mare that knew how to do ranch work while caring for the rider. She just inherently knew the will of her rider and that was why she was such a good kid horse. She was not trained to protect the rider that was just part of her demeanor.

My career had been in the Air Force so Karen had not grown up on a ranch and did not have a lot of riding experience. Lady knew her job was to protect Karen no matter the incident. She protected Karen on several occasions. One I recall vividly. Karen had failed to girt her saddle tightly enough and while leaning too far to one side, the saddle began to slip. Lady came to a complete stop and waited while Karen got off and we came to help put the saddle back in place.

After Karen was away from home, she sold Lady to a neighbor girl who was in need of a good kid-horse. We were told that Lady continued to protect her youthful riders until she became too old for service. The father of the young girl who bought Lady told us of two incidents where Lady protected the rider. One day he noticed that Lady had come to a stop near the pens where the young girl was riding. He went to see why she would not go any further for the girl and found that they had come upon some barbed wire on the ground. Lady's front leg was tangled in the wire and she knew not to try to move lest she hurt herself and the young rider. He told of another occasion when Lady refused to enter a weed patch no matter how the rider urged her. He found that there was a rattlesnake in the weeds and Lady knew the danger and refused to take the rider into the weeds.

While Lady was still here on the ranch, she gave birth to a very nice filly colt that turned out to be an added experience on the ranch. We named the filly Lil. Lil was sort of a hand full from the beginning. A very nice filly but she certainly knew what she wanted and demanded attention to it. I remember one incident that taught me a lesson as well.

Lil was weaned but still running with Lady in a small pasture that we called the horse trap. We had some rams running in that trap and needed to move them into our west pasture. I had saddled Lady to go move the rams. After I saddled and started out in the trap, Barb drove the pickup down to the gate to wait for me and to help me put the rams into the west pasture. Without thinking about the possible consequence, I let Lil follow along on the drive. That was a mistake.

Most young animals learn that they can make their mothers stop by stopping right in the mother's path. Once the mothers stop, the offspring will get in position and start to nurse. Usually the mother will give in and wait until the offspring has had its fill. This way of making the mother stop lasts a lifetime.

This practice is noticed with cows as well. In fact, the cow remembers the tactic from her calf hood days and will use it to stop humans. When we are out amongst the cows in the pickup quite often, a cow will move around in front of the pickup and make us stop. That's her way of letting us know she wants some feed.

Back to the Lil story, we hadn't gone far from the pens until Lil decided to use the "stop tactic" to let me know that she didn't want me riding her mother. She moved so to be in front of Lady and stopped. Lady, being accustomed to minding her daughter, came to a stop. I didn't think much about it and merely spurred Lady to go on. Well, after another attempt or two, Lil had enough. She backed her ears and ran in front of Lady and began to kick with both hind feet. Her ears were back and she was obviously mad. She would run past and kick at my leg in the stirrup. By this time, we were getting into some heavy brush and it wasn't easy to defend myself and Lady from the irate filly. I worked my way over to a fence and got off Lady so that she was between me and Lil. The fence served as protection for my backside. I used rocks, sticks and anything I could find to throw at the silly filly. Finally she got the picture or got tired of the fight and moved away. Lil had calmed down so I remounted, found the rams and took them to the gate where Barb was waiting. Barb had begun to wonder why I was taking so long and was about to search for me when she saw me riding up. Lil had taught me another lesson in the handling of "dumb animals"; leave the yearling filly in the pen next time.

We usually began training our foals while they were very young. We began by teaching them to halter and lead. On an occasion or two, I tried my hand at breaking them to saddle and then to ride. After a few incidences of getting bucked off with the breath knocked out of me, we decided that we would get better horses for future work if we let a professional do the breaking to ride. After breaking Lil to the halter, we arranged for a cowboy from a nearby ranch to take her and teach her to saddle and to train her for ranch work.

This cowboy worked on a large ranch where he would break the horses and then use them on the ranch on a daily basis. In this way, he was able to make a little extra money while taking care of his job on the ranch. A friend warned us about this cowboy. The friend said that this fellow knew how to train horses to work but he didn't gentle them at all. The friend warned us that when we got Lil back that she would be a super cow horse but the first time the rider did something to startle her that the rider might find himself on the ground.

That friend was right. When we got Lil home, the cowboy had been using her daily and she was pretty well worked down. So when I first used her to work cattle, she was super. She knew better than I how to make the cattle do what needed to be done and do it efficiently. Our ranch is small so that there was not a lot of riding to do each day. After she was rested and well fed, I learned right away that the friend's warning about gentleness was very sound.

Lil didn't mind working cattle and would really work well when there was work to do. Just to ride out for reasons unclear to her was not to her liking. One day, Barb and I rode horseback to one of our fields to get a cow and drive her back to the pens. I rode Lil and we got along well traveling the half mile to the field. We rode up to some cows and the cows started to trot away so we needed to move on in order to turn them back. Barb moved out and I kicked Lil to make her keep up. She taught me in a hurry that she didn't like my forcefulness. After about a hundred yards of strong bucking, I departed the saddle. After Barb got over her laughing spell, she rode up to catch Lil and bring her back to me. Lil, probably laughing too, allowed me to remount and took me and the cow back to the barn.

From then on I was a bit skittish about riding Lil. If there was obvious work to do, she was just fine, but you never knew when she would teach that next lesson. We finally sold her to a fellow that prided himself as a horse trainer and trader. I told him about her habits and who had trained her to ride. He was well aware of the cowboy and knew that I was telling him the truth. Later I saw him and asked about Lil. He said that he had taken her into a sandy stream bed and was going to take the "buck" out of her by making buck in the soft sand until she couldn't buck any more. He did, she did, but his training failed to cure the problem and she still does. He said with a grin that he finally just gave her to his "ex wife."

Sadly, horses get older as do the horse owners. As I mentioned earlier, the last horses we were to have on the ranch were Granny and Fancy. When Granny got to be thirty-two years old, she lost most of her teeth and had trouble grazing. We would provide feed daily but her health worsened and the time came when the only humane thing to do was to let a veterinarian put her to permanent sleep. By this time, she was just another member of the family and

Granny In Good Health

the process of euthanasia was difficult to accept. It didn't take many days of witnessing her suffering until we knew the day had come. We had a grave prepared out on the ranch and called the vet to bring the proper medication for euthanasia. He came and the job was quickly over and the burial done. We really appreciated the sympathy note we got in a few days from the veterinarian's wife. She knew that we were suffering the loss of a very close friend.

I remember in Granny and Fancy's younger days they would come to the pen for feed each morning. We would place feed for each horse in the trough where each knew there would be feed for her. Fancy was the boss horse and she would take a couple bites of her feed and then go check Granny's feed to see if it were better than hers. Granny would not argue but merely go to the other trough and continue to eat. Sometimes we would have cows in the pen and they would try to steal food from the horses. If they came to Fancy's trough, Fancy would merely go to Granny's trough and force Granny to go fight the cows. Granny had no fear and was good at cow fighting.

Fancy Mourning

So we were not the only one who suffered from the loss of Granny. Fancy and Granny had been friends here on the ranch for many years. They were the only two horses left and gladly accepted our attention together. Fancy suffered as much as we did from the loss of Granny. We were amazed and saddened even more to witness Fancy suffering the loss of her friend. Fancy would walk slowly around the pen and look here and there for her pal. She even began to leave her food and stand in silence, as if to say "sorry Granny, I should never have taken your food so please come and get some of mine."

Fancy loved attention, both from Granny and from us and our family. Barb decided that it would be better for her if we found her a home with children that would pay attention to her and perhaps use her for mild horse events. We learned that a local school teacher who taught agriculture had three children that were active in 4-H events and offered Fancy to them to be another one of their teachers. They accepted and she was moved to their ranch where she now enjoys much love and attention.

The removal of all horses from the ranch has brought about a sad time. Ranches need horses and now this one is without a horse. Slowly Barb and I are learning to live without the presence of a horse. Surely, we will never forget those good ole days.

New Horse

One of the best purchases we have made was a cube trailer. The trailer holds about a ton of cattle cubes that can be dispensed in piles as it is pulled along in the pasture. The cattle quickly learn to follow along and eat the cubes as they are dispensed. This method of cattle feeding is safe and can be done by anyone who knows how to drive a pickup pulling a trailer. The cube trailer then becomes an excellent "horse" so us old folks can still operate the ranch without a real horse by having the cows follow the cube trailer into the pens.

CHAPTER 2

Cows

As a small, mom and pop type ranch, Barbara and I have close communication with our livestock. Our cows know us and they know our pickup. They know how we normally dress and react differently when we appear before them in unusual clothes. We know each cow, her bloodline and her character. We can readily tell when a cow is not feeling well or has recently given birth to a calf.

We do not operate the ranch as our primary source of income but more as a cattle laboratory. We have been interested in learning about bloodlines and what genetics can bring, not only to beef quality but to the health of the cattle as well. We use an ear-tagging system to identify each cow and her offspring. Then by our observations and weaning weight records we can identify the bloodlines that are superior in our herd. We keep heifer calves from the better-producing cows and alter the bloodline by selective breeding to bulls with differing genetics. Our herd today consists of cows that were raised here on the ranch and are the best of the prior herds.

Ranchers use different methods of identifying individual cows. We use an ear-tag numbering system that allows us to look at a cow today and know her bloodline back several generations. When we tag a heifer that we plan to keep in the herd, we place a number in front of the mother cows ear-tag number. That way, the mother, grandmother, great-grandmother, great-great-grandmother, etc, can all be identified by looking at the heifer's ear-tag. For example, ear-tag 42523 means that this heifer is the fourth one kept from cow 2523. Further, it reveals that the heifer's grandmother was cow 523 and her great-grandmother was cow 23. This system also shows that a cow with ear-tag 32523 is an older sister or half sister to 42523.

We have selected different sire bloodlines to see the effect of crossing those bloodlines. We began our operation with hereford bulls and over the years have used brangus, limosin and angus bulls to provide the cross-breeding that we desired. We have had a few charolais mixed in from neighbor bulls that jumped the fence.

One of the problems with this operation is that one is forced to change bulls more often than a normal operation else there will be an undesirable form of in-breeding. Line-breeding, where the same genetic line is kept in the sire or dame side can be used effectively. Cross-breeding, where the genetic lines are brought in from both the sire and dame can also be used effectively. In-breeding, where the same genetic lines are involved in both sire and dame are acceptable only in certain purebreds.

Care must be taken with sire selection or one may experience severe birthing problems, particularly for the young cows having their first calves. The goal is to increase the weaning weight and quality of the market calves. When we began this program, our calf weights at seven months of age averaged about 400 pounds. This past year our average weaning weight of seven month old steers was well over 600 pounds. Another test of success is the price that the calves bring at market. The goal is to have your calves sell at the upper end of the price range for that weight calves on that market day.

Intelligence is another factor to be considered in the breed selection program. For most operations, one desires cattle that can be easily managed while producing high quality calf for market. Intelligence and character must be considered when selecting what bloodline to keep in your cow herd. We select bloodlines that are easily domesticated so that Barb and I can manage them in the pastures and in the pens.

All ranchers have to "work" their cows. There are times when they must be vaccinated to prevent various diseases. There are times when they must be treated to prevent damage from internal parasites and to control external insects. Usually they do not understand the reason for these nagging interference with their normal lives. They will reluctantly tolerate corralling, herding and loading in a chute to get the job done. In fact, feeding them while in the pen erases a lot of the ire.

A cow that is ailing from some disease or another seems to take a different attitude. Usually when we have a sick cow she will work with amazing agreement with handling. Because they know us and know that we do not mistreat them, they realize that we are trying to help them feel better. They will walk right into the squeeze chute and accept the vaccination or what ever must be done.

Mind you there are times when some will test your skills at cattle handling. Remember they are animals with hierarchical instincts and some times we have to use the snap of a whip to get across that we are the current bosses. Cows have herd instincts and work better while in a group. It's when you must isolate a particular cow that there can be serious handling conflicts and us bosses don't always win.

Cattle are very intelligent and interesting to work if one pays attention. They do not like for you to look them in the eye so they work better if you act like you are ignoring them until you get in position to drive them to where you want them to go. They readily learn handling procedures and you can tell when they know what you want them to do but choose to test you once again. If you are sorting certain ones from the herd, the one you want will quickly recognize that fact and try to hide behind the others. They also learn from each other. If you let one get past you and out of the herd you can be sure that another will try the same tactic. You definitely must be smarter than the cows if you are to be a successful cattleman.

BIRTHING INSTINCTS

Cows will hide their newborn calves by instinct. They will then graze nearby, usually downwind from the calf, so that they can observe any potential threat to the well-being of their offspring. By grazing downwind they can use their sense of smell to detect the presence of a predator. Finding a newborn calf becomes an interesting chore and knowing the instinctive reactions of cows surely helps when trying to find her new calf.

Some cows seem to be proud of their new calves and are ready to take you to see. When you come upon one of these proud mothers, she will look toward the calf and start in that direction so that you may merely follow along. Once at the calf, she will talk gently to the calf to assure it that she is near and not to be afraid. It is always a joy to witness the pride of these mothers and to see their reaction when you brag about the quality of her calf. The cow will make a low, motherly, sound that is obviously calming to the calf. In fact, this cow-calf talk is what the calf first hears from its mother as she begins to clean away the birth sack as she witnesses those first movements of a live offspring from a successful birth. She will continue that conversation until the calf gains strength, staggers to its feet, gains its balance and begins to nurse. I am always pleasantly amazed at the instinct of these newborns; an instinct that is so necessary for survival of their species. In humans, God gave that instinct of survival to the mothers rather than to the babies.

I remember one experience that turned out to be real sad. We had noticed one cow that had calved but did not have the calf with her. It was a busy day so Barb drove over to check on the cow. She found the new mother who obviously wanted Barb to follow her to the newborn calf. Barb followed as the cow led the way. They went a lot farther than is usual for a new mother to be away from her baby. Barb was beginning to think that she was taking her on one of those wild goose chases and that her calf was probably in the other direction. They came to a small water pond formed in the bed of a creek that was no longer flowing water. Then Barb saw the problem. There in the water was the baby calf. The cow had chosen a birth spot that was too close to the water and, apparently, as the newborn began to find its balance, it had stumbled into the water and drowned.

Barb quickly came to find me and she was very distraught. We drove back to find the cow still waiting for us to help. It was a sad experience to find the dead calf but knowing that the mother expected us to help save her baby made us even more sad. There was nothing I could do but drag the calf from the water and leave a grieving mother to say goodbye to her baby.

This mother wanted our help so she took us to her calf. Other cows with newborn calves are not so helpful and will use various strategies to lead you astray. When you first encounter one of these deceiving new mothers, she will nearly always take one "quick look" in the direction of the calf. Experience teaches you to watch for that "first look" because that will usually be the only clue you will have as to where the calf is located. Many times, after that "first glance", she will try to lead you in the opposite direction. I've wasted many hours trying to find a calf by going with the mother as she led me away from the calf.

These fruitless jaunts quickly teach you to think before you follow. Did I miss that "first look" and she knows that I was not very observant? Is she taking me down wind from where I first encountered her? Most likely the calf is upwind from where the mother is grazing. Is she looking under every shrub and then going on to the next? A cow that does that knows where the calf is hidden and fruitless searching is a ploy to lead you astray.

If I become suspicious, I will go in the opposite direction rather than continue to follow, imitate that mother-calf talk and watch her reaction. Quite often they will give in and help you find the calf. Another tactic they sometimes use is to pass near their hidden calf and not even look in its direction because through peripheral vision she could see the calf was still there. I have been led right past a hidden calf only to discover that fact after considerable time spent on fruitless ventures.

TWIN CALVES

As is true of most animals, cows sometimes have twin calves and their protective instincts can be even more confusing. I suspect that ranchers have more incidences of twin calves than they know. Having two babies to hide and protect must be very confusing to the bovine species and it can be confusing for the rancher as well. Sometimes you come upon a cow that has just had twins and she will have both of them near her, but more often that is not the case. You learn about the "other calf" by accident. For example, you may find a cow with a heifer calf and then later find her mothering a bull calf with the same parental pride as she demonstrated when you saw her with the heifer.

Cows will nearly always keep the twins separated. She may keep them separated by a mile or so as her instincts tell her not to risk both of them to some peril. She will let one of the twins nurse and bed it down. Later she will graze her way to the other twin and let it nurse. The very young calves instinctively stay where mother put them even through they may get hungry or scared. It is easy then for the rancher to interfere with this twin-mothering chore. For example, he may move the herd to a different pasture and thereby separate the mother from her other calf that he did not know was out there.

I recall one incident where the mother took me to a calf that was dead. I assumed that the calf had died at birth and that we would just be short one more calf at sale time. Then later I found her nursing a calf that was obviously hers. The dead calf was an unfortunate twin. She had taken me to it trusting that I would help bring the calf to life. It's sad to be a failure at times like that.

Another time, I happened upon a cow with twin calves at her side. I knew that they were hers because she was the only cow in the small pasture at the time. Later I went to check on them only to find her with just one calf. After much discussion and considerable looking, I finally came upon the other calf bedded down under a shrub. The mother cow never let on that she knew it was there so I drove her over to the calf and then she began to mother it just as she had been doing with the other twin. These twins were of different sex so we ear-tagged the heifer with the letter A and the bull with the letter B. We put her in with several other cows that had calved. Sometimes we would find her with A and sometimes with B. It was several weeks before we found A and B with their mother at the same time. The following year that cow had twin calves again and we tagged them C and D.

We learned that you should not interfere with the instinct to keep the twins separated. I recall a couple of times that we caused a cow to reject one of the twins because we confused her. Once we found a cow with twins only to later find her several miles away with only one of the calves. We went back to the area where we had first found her with both calves and found the other twin. Stupidly we, being intelligent humans, thought we knew best so we caught it and took it over to the mother. She was totally confused and refused to take it as hers. We bottle fed the calf until it could take care of its own nourishment. Another time we knew of a set of twins and their location. We found the mother and one twin and drove them toward the pens on a path that would go by the other twin. When we got to the other twin, the mother did not acknowledge it so we forced it to go along with us. Once in the pen, the second twin would try to nurse and the mother would kick it away. Obviously, we had confused her and we ended up with another orphaned calf. Fortunately this time we had a cow that lost her calf at birth and we were able to get her to adopt the orphaned twin and raise it as her own.

Twin calves are often smaller than single births and there are genetic differences that may affect their ability to procreate. Most ranchers do not want cows to birth twins. Our experience has been that the cow produced more dollars from the sale of twin calves than we would have gotten from the sale of a single birth. On the other hand, if you lose one of the twins because of birth defects or human error, the sale of the smaller surviving twin will not bring as much money as she would have produced from a single birth. We do not think it would be wise to breed for a herd that would genetically produce a large number of twin births.

PARENTAL HERD INSTINCTS

Cows with calves at side will tend to stay together and sometimes form a herd separate from the cows that have not yet calved. There are some interesting things we have noticed about the instincts of cow mothers. A cow does not seem to discipline her own calf but leaves that to the other cows. If a calf tries to nurse a cow that is not its mother, the mother does not try to stop it but the cow being violated will surely let it know that she is not its mother. After a strong kick or two it gets the message that mother is the only one that will tolerate being nursed. Calf discipline becomes not what one is taught so much as what one may get away with successfully. A human trait?

Cattle do form nurseries. Frequently, one cow will stay with a group of calves while the other mothers go off to graze. We have never figured out how they decide who the nurse mother will be at the time. It will be a different cow that stays with the nursery at different times. You can be sure that the chosen nurse cow is very faithful while it is her turn and will diligently protect the calves in the nursery. After grazing awhile, the mothers will come back to feed their calves, a routine that seems to be a normal "dumb animal" thing to do. Our intelligent human thought is "why did she serve in charge of the nursery without pay and how could those mother cows trust that she would protect their calves?"

ONE FINE AGGIE COW

Cows definitely have some traits that show their intelligence. This Aggie cow definitely demonstrated her skill of communication.

Texans have always argued over which university in Texas best represents agriculture. The University of Texas has a longhorn steer as their mascot although their contribution to agriculture is questionable. But the "A" in Texas A&M University stands for "agriculture" and most Texans know the truth. Texas A&M University is the best.

Cows know the truth too and some definitely have favorite universities. We had one cow that was a strong supporter of Texas A&M University. She was a true Aggie Cow.

I don't remember exactly why but we had this aggie cow in the pen by herself. I had found it necessary to go in the pen with her several times. Each time she would warn me to stay away by squaring off while pawing the dirt and snorting. I would act pretty aggie macho and let her think I was ignoring her while getting out of the pen as fast as my reason for being there allowed.

Students at most universities have neat hand signals that symbolize their schools. The Aggies use a closed fist with the thumb up that symbolizes the "gig em" Aggies slogan. The students at the University of Texas, tea-sippers, extend the index and pinkie fingers with the middle fingers folded down to symbolize their "hook em" horns slogan. The "gig em" and "hook em" hand signals have a definite affect on the attitude and demeanor of the students of the opposing university. These hand signals have been known to cause some fist-fights.

That's how I came to know that the cow we had in the pen was an aggie cow. I found it necessary to go into the pen with her one last time. She greeted me as she had been doing, squared off, pawing the dirt and snorting. By now I was sorta fed up with her warnings and took them as nothing more than false threats. Then I made a big mistake. I showed her the "hook em" horn sign and waggled it in her face. That really stirred up that aggie spirit and she came at me with the acceleration of a fighting bull attacking a matador. Alertly I managed to get back though the gate just before she slammed into it.

Some cows definitely know the universities that serve them best. She was a true blue-ah maroon Aggie Cow.

TELEPATHIC

Cows are definitely telepathic. I don't know whether the telepathic signal begins with the cow and we react or begins with humans and the cows react. I think that some cows do read our minds and try their best to interfere with our lives.

The cows can be grazing near the pens for days but on the morning we plan to put them in the pen they will be at the back of the pasture. We wonder how they know that we decided to put them in the pen on that day. Imagine how confused they must be to migrate to the back of the pasture on that morning when the grazing was better near the pens.

We have tried every approach to the problem. We have tried to fool them by sending a mental telepathic message that we will not need to pen them on that day when it is planned. We have tried to send the penning signal for several days in a row to try to fool them but nothing works. Whatever the day, they will have traveled to the backside of the pasture overnight. I seem to be getting the message that they would gladly be in the pen on the proper morning if I would leave the gate open and plenty of feed in the troughs.

Another example of cow telepathy occurred recently. We had been invited to a reception at the Governor's mansion on Saturday afternoon. A heifer had been showing natures signal to calve for several days but wouldn't go into serious labor. Guess what? Somehow she sensed that we planned to go to the Governor's bash and chose to go into labor on that Saturday morning. Nothing to do but wait her out and help as necessary. I hope the Governor understood that the cows take precedence over politics.

AGRICULTURE EDUCATION

Our son, Ken, took an early interest in cattle. During his high school years, he participated in vocational agriculture programs where he learned cattle blood lines and was active in the calf showing part of the program. We tried to help by providing good quality calves for him to feed, train and show. We were especially proud when his heifer, Dorie, won Reserve Champion Heifer at the Houston Texas Livestock Show. He brought Dorie back here to the ranch and she added excellent genetics to our blood lines.

The excellent agriculture programs also taught him to communicate effectively through writing skills and, as a member of the debate team, he learned how to communicate orally. Ken went on to earn a Masters Degree in Animal Science from Texas A&M University and with ample on the job training became Vice President of Beef Procurement for Excel Corporation, Wichita, Kansas.

Much to our pride, Ken passed on his learning skills to our grandsons, Justin and Michael. Justin is finishing his Masters in Sports Management at Texas A&M University and Michael is on his way to degrees in Engineering at the University of Kansas.

Let me further proudly digress to state that Karen's daughters, Diana and Donna, are also furthering education. Diana has her degree in elementary education and is teaching elementary students in Canyon, Texas. Donna has her degree in psychology and plans further degrees at the A&M university in Canyon, Texas.

CHAPTER 3

Sheep

As mentioned earlier, we tend to think of sheep as the bill payers rather than intelligent animals on the ranch. They certainly know how to make the rancher pay for the expected profit. They must be fed, drenched, vaccinated and sheared. Each animal has to be individually treated by hand. They must be penned and then caught one by one whether to be medicated or sheared. They don't like to be caught and will use all their energy to prevent handling. They will use their head to attack you or they will use their hind legs to kick you. The rams are even stronger and some have horns that they are willing to use. It takes a strong ranch hand to work the sheep.

Sheep are intelligent enough to know that they are prime food for predators. Their main defense is herd instinct. They move about in herds and they sleep in herds. They are ever watchful for predators and will react to potential dangers as one body. When it comes time to bring them into the pens, that desire to stay together can be helpful.

We have noticed a definite intelligence of these dumb sheep as related in this event about one of our rams.

Early in 1992, Barb and I had gone to the West Pasture to feed the cows. We had gone down the north side to find the cattle near the water trough. After we called them together and put out their feed, we started back to the barn by way of the south route. This is usual practice, allowing us to see more of the pasture and a better chance to find any animal that might be in trouble. The sheep were still in the West Pasture at that time and one never knows what dire trouble some ovine animal might encounter.

Bluff Creek cuts across the eastern part of the West Pasture and along the creek Barbara's Dad had dug three earthen reservoirs (tanks) to store water for the livestock. The trees had been left undisturbed around the banks. As we neared the Bluff Creek tanks, the road veered right around the up-creek extension of the first of the tanks. Recent rains had filled all of the tanks causing water to stand well up the creek. Suddenly Barb said, "Look at that Ram." I looked up the road and didn't see anything.

"Over there in the water", she said anxiously.

There, standing in water up to its belly, was one of our larger blackfaced rams. He was uneasy, moving right then left, facing the bank near a couple of dead trees. Every now and then he would sip a little water and let it dribble from his mouth. His attitude was almost like he was in a coma or running a high fever.

She had stopped the pickup and we started toward the ram. Barb said, "shouldn't we get a rope?" As is usual when you need something you don't have it with you, but then a pickup will only hold so much and you can't be prepared for every catastrophe.

"We don't have a rope with us", I retorted.

She went back to the pickup and began to drag out some nylon cord that we carry to tie things down in the truck bed on occasion. I went on over near the ram. When I walked up he turned away as if he planned to try to escape me and cross to the other side.

We keep our sheep gentle by feeding them now and then. It's a good idea to keep the rams gentle, if you can, because they are big enough to hurt you badly. Most ranchers have been hit in the back by a ram and quickly learn that they are a definite hazard. I had these rams gentle enough so that they would stand for me to drench them without having to tie them down.

As he turned, I called to him in the tone of voice I use when we feed. He did recognize the call although his demeanor didn't show it. I kept talking to him trying to figure out what to do. The water was too cold for me to wade in and even if I had, I would not have been able to lift that large ram out of the water.

Barb brought the nylon cord to me and I began to maneuver as close to him as I could hoping to get the makeshift rope around him. I figured if I put the rope around his head, he would merely set back and make matters worse. So my plan was to make a large loop in the cord and try to pitch it over his entire body so to put pressure on his rump. Maybe that pressure from behind would cause him to climb up the bank and out of the water.

Anyone who has worked with animals on a ranch will learn that most animals, although we call them dumb, know when you are trying to help them rather than hurt them, particularly when you have handled them gently in the past and they have benefitted from the experience. I think that this ram began to realize that we wanted to help him. Much to our surprise and delight, all of a sudden, without my having a chance to try the cord-over-the-rump idea, he began to climb out on to the bank near to where I was standing. I merely guided him and he walked the 30 feet or so to the back of the pickup with only mild coaxing.

It took both of us and a lot of grunting to lift him in to the back of the pickup. It was more like we rolled him in rather than lifted him in. I rode in the back with him as Barb drove us to the barn.

It would help if ranchers were trained veterinarians, but most of us have to gain that training just from the experience of working with animals every day. Our layman examination did not reveal any wound or visible problem. To be safe, I cautioned Barb not to get his saliva on us as he might have rabies, a precaution that I really didn't need to offer since she knows as much as I do about animal husbandry. The ram was very drawn looking and he had a nasal discharge, but then nine and a half out of ten sheep will have a dirty nose. He was a bit unsteady, breathing with difficulty but not wanting to lie down. I went to the refrigerator we keep at the barn for medications and came up with some penicillin and Ivomec. I gave him a large injection of penicillin and the correct amount of Ivomec in case he had a bad case of stomach worms. After the medication we made sure that he had some fresh water and sweet feed and left him to his fate. There was nothing more that we could do but pray and hope.

He was no better the next day. In fact I was sure he would soon die. His breathing was very labored, having to draw hard to get air in and then harder to get it back out. He stood around with his nose draining and paying no attention to anything around him. Even when I gave him another shot of penicillin he didn't react.

On the third day he was even weaker and could hardly stand. We were sure that he did not have rabies, so we cleaned his nose and I forced (drenched) some cold water into his stomach. He had not eaten nor drank since we brought him to the barn. We put out fresh water and a pail of oats for him to eat.

Apparently the getting some cold water into his stomach helped because he was much better on the fourth day. He had eaten some of the oats and was feeling well enough to bleat and come toward me. His breathing was much less labored. We kept up the penicillin treatment.

His recovery continued and we kept him around the pen for several more days. Soon he was following us around the pens asking for more to eat. We knew then that he was ready to go back out with the other rams.

It seems at times that sheep are just looking for a place to die. They are very vulnerable to diseases, especially respiratory diseases. Usually it is a foregone conclusion when one is sick that it will soon die no matter the husbandry. Occasionally, as in this case, you do get one to recover and it makes all the failures quickly forgotten.

CHAPTER 4

Sheep Dogs

Sheep can be fearful of dogs but they are intelligent enough to learn to accept the protection of sheep dogs. Their herd instincts work out perfectly for the sheep dogs so as soon as they learn that the dogs are trained to protect them, sheep settle down to enjoy the added security. We had to learn for ourselves that trained dogs can be helpful with the care of sheep. Our experience began on a cold night much to our surprise.

It was about four on a cold January morning when we were awakened by sheep bleating and a dog barking. We turned on the outside lights and I grabbed a flashlight and went out to investigate. The sheep were well acquainted with our house and our habits since they were grazing in the pasture surrounding the house. They came running to the light in hopes of getting away from the dog. They were not accustomed to dogs offering protection, only dogs that were a danger, so they were coming to us for help.

I shined the light to the outskirts of the herd and there was a large white dog. The dog was circling the herd wondering why the sheep were scared of her. She had been trained to be part of the herd and did not understand their thinking that she was their enemy.

I recognized that it was of the Pyrenese breed of sheep dogs. The only ones in our area belonged to the Bratton ranch that joined our ranch on the south. We called the Brattons next morning and they confirmed that it was one of their female sheep dogs. They had recently sold all of their sheep and had put the dog with their goats. As often happens, when a dog is trained for one species it will not perform guard duties for another species. This dog had gone looking for more sheep to guard and had found our herd. The Brattons, being good neighbors, suggested that we allow her to stay with our sheep. We named her Brat.

Our sheep quickly got accustomed to Brat being around and settled down to their normal herd habits. We made sure that Brat got some dog food each day by placing it in an area where we were feeding the sheep. Brat would not come to us but she was always

Brat

there with the sheep. We had been hearing coyotes at night and were glad to have her there to guard the herd.

When our grand kids were young and visited quite often, we put up a swing set and a slide in a play yard. We didn't want the play equipment in our yard because of the problems it would cause when mowing. The play yard was just outside our yard fence and it was fenced to keep the ranch animals out.

The play yard turned out to be an excellent place to put orphaned lambs that required bottle feeding. We put a small house in the pen to give the lambs some shelter. It was easy to take a bottle out to the fence and insert it into a bottle holder and let the orphan lambs nurse.

Lambs in Pen

Brat would come to the house during the day rather than staying with the herd and she became accustomed to the three lambs that we had in the play yard at that time. Brat would go out before dusk, find the sheep herd and spend the night on guard duty. Early next morning she would return to the shades near the play yard and settle down to sleep. The orphaned lambs got accustomed to her habit. We noticed that she would go to the pen and check their well being several times during the day.

The orphaned lambs grew and soon it became time to let them wander outside the play yard. Brat would watch them and the lambs expected her to be there. One day as they got bolder, the herd of sheep came near by and the lambs joined the herd. We had not noticed them being gone but Brat was well aware that they had left the play yard and had gone with the herd. Before dusk that day, Brat went to the herd and somehow got those lambs to follow her back to the play yard. Barbara was outside at the time and witnessed as she brought the lambs back to the play yard. Brat waited until the lambs went safely inside and then she casually wandered away to rejoin the herd for the night.

The lambs were happy to be back in their home environment. This was a clear case of two species of "dumb animals" being intelligent enough to realize that home was the best security for those orphaned lambs.

Brat was a faithful companion of the herd and when we worked the sheep or moved them to another pasture, Brat went along. She quickly taught us the value of sheep dogs.

Along about August, we noticed that Brat was not with the herd. We called the Brattons to see if she had gone back home and they sadly reported that they had not seen her. We never did know what happened to Brat but she never came back to either ranch. We only hope that she found another sheep herd to adopt and that nothing bad happened to her.

Brat had taught us that sheep dogs could provide the much needed protection from predators. She was always there with the herd and could sense any approaching animal that might harm the herd. She was constantly on patrol and would not hesitate to go on the attack. Usually her bark and advance toward the danger was enough to scare off the intruder.

We decided to replace her and began the search for another dog. We located a ranch that raised and trained Anatolian sheep dogs and bought a young male that had been running with a herd of sheep on that ranch. We brought that dog to our ranch and let him spend a day or so in the pen with a few of our sheep so to become bonded to our herd. This dog also proved to be an excellent guardian for our herd.

It was very odd, sad and unexplainable but this dog disappeared in August as had Brat. We never found him and no one at the neighbor ranches ever saw him.

Our next experience with sheep dogs taught us that you must be alert or you may buy the problems that a previous owner was more than willing to sell you. We went through two such sets of dogs.

The first pair that we bought were older dogs and we could never get them to settle with our herd. They seemed to have gotten spoiled to roaming the neighborhood at will. We took them back and got a young pair. We thought that we could start them out in the pen with some of our ewes and that would train them to stay with our herd.

This pair we named Sandy and Candy. The herd training was successful in that they were faithful to stay with the herd. We began to find some lamb kills in the mornings when we went out to check on the herd. We had been hearing coyotes call at night but the kills were not all typical of coyotes. Coyotes usually will kill and then open the abdomen and clean out the parts that they find most edible, usually the liver and lungs. Coyotes will seldom eat much of the meatier pieces of the carcasses.

We were noticing that some of the carcasses would be well-eaten and became suspicious that the kills might not all be from coyotes. I began to watch. Apparently the young dogs would find the left-over carcasses and eat on them. They decided that they would not wait for the coyote kills and take the lambs at will. Sure enough, I happened on the young dogs as they were eating away on a fresh lamb kill that I was certain the lamb had fallen to their prey. Once dogs become killers themselves there is no hope of correcting a sad situation. Since we had prior experience of buying someone else bad dogs and not wanting to treat some potential buyer to that same experience, I sadly took the matter in hand and eliminated the problem forever.

The coyote problem became worse without dogs. Coyotes were coming to the ranch from the north and the south. One night we heard a coyote fight as the coyotes from the differing locations came and were attacking the sheep herd at the same time. We would go out in the morning to find several kills. I remember one morning when we found a lamb victim that had survived the attack and was lamely following its mother. Usually the coyotes will attack the lamb by grabbing it's throat and causing suffocation, but this attack was different. Apparently, the coyote had grabbed the lamb

Coyote Damaged Lamb

by the right rear leg and had literally torn away the flesh to expose the bone. The poor lamb was hobbling along following it's mother and you could see the bones in the leg. We drove them to the pen and did our best to bandage the lamb's leg. It was no use, the damage was too severe and the lamb died of massive infection.

Now we were out of dogs again and the dog solution to predator problems in our sheep was becoming rather expensive. Losing lambs is also expensive and so we decided to try one more time. This time we decided to get a male and female pair of Pyrenese pups as soon as they were weaned.

We had decided that it might be interesting to buy a male and female pair of Pyrenese dogs and produce our own dogs as we might need them. We found a reputable Pyrenese sheep dog farm and went to see the pups. There five of them all white, fluffy, waddling puppies curious about everything, pouncing on each other in good humor. There was only one male pup so it got chosen right away. As we watched the four females, one was far more bold and curious than the others and Barb decided that she would be the one. We settled the bill with the owner and put the pups in a pen in the back of the pickup. We wrapped the pen so that the pups would be protected from the wind on the way home.

Since everything on the ranch has to have a name, we discussed what to name them as we drove home. Barb suggested that we name them for the grand kids but there were four grand kids and only two pups.

DeeDee and Jim

Barb suggested that since the grand daughters were named Diana and Donna that we could use the D from each and she came up with the name of DeeDee for the female. I had to follow suit and so I suggested that since the grand sons names were Justin and Michael, why not J for Justin and M for Michael and when put together the J and M would sound like Jim. So we proceeded to the ranch with new sheep guardians named DeeDee and Jim.

We penned a few ewes and their lambs to use in the training procedure. We kept DeeDee and Jim in the pen with that set of sheep for a week or so to allow them to get to know each other. It wasn't long before the sheep and dogs were at home with each other. The pups and the lambs would play as if they belonged to the same species. We didn't know whether the dogs thought they were sheep or the sheep thought they were dogs but the bonding was successful. When we turned that training group out with the larger herd, it took a short time for the untrained part of the herd to accept the smaller herd and the dogs, but soon they were as one.

DeeDee and Jim Grown

DeeDee and Jim were super. It didn't matter in which pasture we had the sheep, the dogs were there as part of the herd and did an excellent job of protecting the sheep from predators. After the appropriate length of time and with natures blessings, DeeDee became pregnant with Jim's pups and we waited for the day when she would bring some new Pyrenese pups into our world. One day we noticed that Jim was with the herd but DeeDee was not around, while that was unusual we knew the reason.

The next day when we visited the herd, DeeDee came in and we noticed the direction from which she came. It was obvious that she had given birth. We started to search in the direction that she came from and she willingly helped us go to the makeshift home she had built under some brush. She and Jim were proud to show us the seven babies that were there in a nest. They were only a day or two old and had not yet been able to open their eyes.

This birthplace was a long way from our house and it was winter. We borrowed a deer blind that the hunters had built from half sheets of plywood and moved it near the birthplace. The "dog house" provided some warmth and comfort for DeeDee and the pups. Jim was not allowed to be inside with the pups.

In a day or so two of the pups had died from unknown causes but that fact created concern. We decided to move the dog house to a location near our house so that we could see the family more often and perhaps prevent additional deaths. We moved the sheep herd along with the dogs so that all would be more comfortable with the location.

Dee and the Pups

This arrangement worked nicely and we thoroughly enjoyed watching the family grow. Jim would spend time with the sheep and occasionally with DeeDee and the pups. It was amazing how fast they grew. It was no time before they were playing, biting each others ears and generally being big pups.

The pen was fixed so that Jim and DeeDee could come and go by jumping over a gate. The pups could not get out of the pen. Jim and DeeDee would go in search of food and gradually teach the pups what they could eat to supplement mother's milk. The parents would bring in an animal carcass and the pups would argue over the right to chew on the carcass.

One day we noticed that Jim had been out with the sheep and a small lamb had followed him back. The Pyrenese are white like the sheep and the lamb had apparently become confused and thought Jim was its mother. Jim was also confused as he didn't know what to

Lamb Nursing Dee

do about the situation. DeeDee solved the problem. She considered the lamb to be no different than her pups and when she fed the pups, the lamb merely joined in and nursed. That was quite a sight for us. Who would have thought that a dog would have allowed a lamb to nurse along with her pups?

Nursing done, the pups set out to play and the lamb got introduced to the rough play of the pups. They would grab the lamb by the ears and wrestle it to the ground. The lamb found that behavior unbearable but its only defense was to bleat and run. Bleating and running was no defense, the pups found that running merely added to their play.

While it was an unusual experience, we knew that we must interfere with this joining of the species and save the poor lamb from the play of the pups. Usually, when a lamb has been separated from the herd, it is impossible to find the mother that will take it back. We had no choice but to try and so we took the lamb on a search for its mom. This time it worked out. We found a small group of sheep that

Rough Play For Gentle Lamb

were away from the herd. We put the lamb out near by the group and it began to bleat. Much to our surprise and pleasure, a ewe came from the group and claimed the lamb. The lamb acknowledged the mother by starting to nurse.

The pups grew and we let them run freely with Jim and DeeDee. At first they would stay near their pen and wait for their parents to bring home the food. Eventually, one or two of the braver ones began to go out with their parents and experience the life of guard duty.

It wasn't long before we realized that seven large dogs would be more than we needed, so we advertised "Pyrenise sheep dogs for sale". All we needed to do was get the potential customer out to see and the sale was successful. It was no time until the last pup was sold and gone.

Jim found the pups being gone to be no more than back to normal, but that was not the attitude of DeeDee. She looked at the loss of her pups as being a violation of trust between her and us as her guardian. Within a few days after the last pup left the ranch, DeeDee disappeared. We found her at the Bratton ranch and brought her back home only to have her go back to Brattons to emphasize her anger. Jim would stay at home and tend the sheep.

By this time, the sheep industry was beginning to experience problems. The market for wool was decreasing and the cost to shear the sheep was increasing. Coupled with the continuing loss of lambs to coyotes, the time had come for us to get out of the sheep business. We were getting older and manhandling the sheep was no longer worth the small profit. We sold our sheep herd and since DeeDee was already at the Brattons, we took Jim over and gave them both to the Brattons. Turn about was fair play, our first experience with sheep dogs came from the Brattons and now it was only fair to return the favor by giving them our good dogs.

CHAPTER 5

Goats

There are several kinds of goats of interest to ranchers, some are raised for the mohair and some for the meat product. We have raised both kinds but found that the hair goats were too much trouble. They had to be handled much as sheep requiring frequent medication, one-by-one handling, shearing and marketing of the mohair.

We preferred raising Spanish goats. The Spanish word for kid goat is "cabrito". Around Texas, cabrito means barbecued kid goat and it is a delicacy that makes the raising of Spanish Goats a profitable business. The problem the rancher faces with goats is predators. Coyotes love kid goat and the nannies like to go into the roughest part of the pasture to birth their kids. That is what the coyote wants—easy prey—the nanny placed the young tasty kid right onto the coyote's table.

The young Spanish goat kids are really adorable animals. They are of all different colors and are very playful. One cannot help but stop and watch them play. Along with the playfulness comes a lot of mischief. They get into everything available to them. They love to climb upon whatever and will chew on everything.

Before we built our house out on the ranch, we lived in town in the house where Barb grew up. We would go to the ranch every day to take care of the animals and perform the necessary functions for the day. One such day, we noticed a newly born kid goat near one of the stock ponds. We were running a herd of Spanish goats at that time and seeing a young kid was not unusual. We merely assumed that the mother had wondered off with the herd and would come back for the kid.

The next day we were especially watchful to make sure that our prior day assumption was factual. It wasn't and the baby goat was still in the same area. It was so small and weak that we were able to catch it and examine it for wounds, thinking maybe it couldn't follow the mother because of physical disability. There were no wounds and we decided that the mother had abandoned the baby for some reason and wouldn't return. We decided to take it with us to town because we are basically a couple of softies.

We took the kid goat into town and prepared a place in the back yard so that it would be safe and secure. We named her Nancy Goat. Barb prepared a bottle of milk for Nancy Goat and we began an artificial mothering ritual. Morning, noon and night Nancy got her feeding and began gaining weight. Milk was the thing she needed and she was more than happy to have us be her parents. Soon she didn't even know that she was a goat. Humans were her world.

Nancy was solid white with dark ears, cute and very playful. Before long we would let her roam the yard when we were at home and she loved the freedom. If we were sitting out in the evening, she would come jump in someone's lap and thoroughly enjoy the attention. We had to be watchful because you never knew when she would come running, twisting and turning. You could never tell when that particular romp would end up with a jump in your lap.

Nancy grew quickly and we began to take her with us to the ranch. We would let her out at the barn and she would just make herself at home with the other animals around the barn. After a while as she became accustomed to the barn surroundings so to be safe there, we started leaving her out there on the ranch. By this time she was past needing the bottle and could make her own way eating the grass and browse near the barn. She would always greet us in her cheerful and playful spirit each time we arrived. Quite often she would jump up on the hood of the pickup so she could see us inside the truck. As soon as we got out she was ready to play.

That is what made one particular day so sad. We got to the barn and Nancy didn't greet us as we had expected. We assumed that she had wandered off and would return even though that had never happened before. We went ahead with our chores and kept an eye out for Nancy Goat. She didn't show.

Many years before Barbara's dad had built a rock reservoir near the barn to hold the water from a well. The reservoir was built with thick rock walls and a concrete bottom. It was round, about fifty feet in diameter and the walls were about five feet high. The walls were a foot thick. When full of water it was a super swimming pool and we kids enjoyed swimming in the cool, clean well water on a hot day. The thick walls made it easy to dive in from anywhere and a bathing suit was only necessary if there were bystanders..

When Nancy didn't show up to greet us on that one particular day, even after we had been around for awhile, we became very concerned for her safety and began to search for her. The ranch does have predators of differing kinds, coyotes, bobcats and even large racoons can decide to have goat for supper. For some reason, Barb decided to look into the reservoir and I can still hear her scream. Apparently, Nancy, being a playful goat kid, had learned to jump on a limb of a mesquite tree near the side of the reservoir and from there onto the thick walls surrounding the pool of water. We had seen her up on the wall and thought nothing about her safety. We don't know whether she decided to get a drink and slipped, go for a swim or what but there, floating lifelessly in the pool of water, was our Nancy Goat. For whatever reason she got in the water, the water level was low and apparently the walls were too high for her to jump, get a secure foot-hold and lift herself safely from the water. She had become too exhausted to stay afloat and had drowned. I'm sure she called for us to help but we were not there when she needed us.

Nancy Goat was as much a part of our lives as had been the various dogs and cats that were our pets while the kids were young. With so many really cute animals around a ranch, softies like Barb and myself have to learn not to become too attached to particular ones lest we then suffer their loss. Nature just doesn't care whether some human might mourn the fate of a particular animal.

Nancy Goat

This story that I wrote
About our Nancy Goat
Was not meant to be sad
Although the ending's bad.

Her life brought us joy
And even death cannot destroy
The pleasure that she brought.
No reason to be distraught.

That's just part of life
There's joy and there's strife
All part of God's plan.
Enjoy life while you can.

Goats are useful in that they eat browse and help control the underbrush in a pasture. In the areas where goats are the primary animal stocked on a ranch, the area will be void of underbrush with only the larger trees left standing. We had sold our goats and soon the underbrush began to grow.

We needed some help with underbrush so we bought a small herd of Spanish goats to help us clean up the pasture. The pasture we wanted them to work was the one where our house is located. I guess the prior owners of those goats had domesticated them. They were rather tame and didn't mind coming around the barn or our house. They got into everything — not just the kids but the older goats as well. In the winter, they soon found that our driveway would warm up during the day and give off that heat in the colder night time. That made a good place for them to spend the night.

It wouldn't have been so bad if they had merely slept there. Goats don't respect each other nor their human neighbors. Some are constantly jumping, bleating, banging into the yard fence or garage door, while others tried to sleep. To tell the truth, goats also stink. They leave deposits that smell badly and are not in the habit of cleaning up after themselves.

We tried to discourage them from spending the night on the driveway. We would go out noisily and run them away, hoping that they would get the message that we didn't want them there. They would run about 50 feet away, turn and stomp their feet and snort at us. As soon as we would go back into the house, back to the driveway they would come.

Not wanting to fence ourselves away from the garage, I decided to put up a temporary fence to discourage their spending the night on the driveway. It didn't work. They would climb up on the fence and push it over and come right on in.

This poem tells of our experience and the ultimate end to the problem.

CABRITO, OLE

I've heard it said and I quote,
"Happy the day I bought a boat,
But when the truth be told,
Happier still the day it sold."
But the person who made that quote
Never bought then **sold a goat**.

Out on the ranch the grass is lush
And there is lots of brush.
Goats eat brush, so to hedge our bets
We added some goats to our assets.
The Spanish kind, with colored coat.
Several nannies and one billy goat.

We'll make mistakes, we realize,
So this one was no surprise.
Who would guess, but now we note,
That nothing's safe from a darn goat.

To them there is no end
Up, over, under and in.

They got in the barn and there began
To turn over every bucket and can.
They climbed on the tractor and the truck.
And some in the fence got stuck.
They ate everything they could see.
They got in the orchard and ate a tree,

But the worst, I've yet to say,
They found our driveway one day.
My oh my what a lark!
They got in where our truck we park.
And on our driveway spent a spell;
What they left was more than smell!

So the happy day is on the morrow
As off to market; there's no sorrow.
Except for maybe the coyote.
He really did like kid goat.
For Barb and me, we can only say,
"Good eating folks; cabrito, ole!"!

Dumb does not describe the Spanish goat. Dumb is a better
description of the person who tries to keep them in the pasture that
includes the house.

CHAPTER 6

Wildlife

Nature requires that various species of wildlife must live on the ranch land. The ranch owner has no right to object to that fact of nature and most of us learn to cope with the various wild species that live on the ranch. We know that each animal or bird has its proper place and while we may not be sure what their purpose may be we must accustom ourselves to their existence. After all, according to the nature lovers, we are the ones who are intruding on the habitats of the wildlife by bringing agricultural livestock onto the premises. I guess they may be right but some of those folks might not have plenty to eat if we ranchers did not produce their food out here on the farms and ranches.

We find all of the animals to be attractive in their own way and we don't try to control them unless they become a repetitious problem to our normal operations. Some of the animals are down right cute and fun to be around. I have trouble seeing anything cute about a coyote, even a coyote pup, but raccoons are really beautiful. Opossums are rather ugly but armadillos are pretty, especially the young ones. I startled a group of four young armadillos once and they ran to cuddle my boot as if it was their mother. I sort of liked that experience.

We have lots of snakes on the ranch. Some are helpful in controlling the rats, mice, insects and similar pests. We don't mind the non-poisonous species being around so long as they don't live around the house. Non-poisonous snakes may not be harmful to humans but when they show up unexpectedly, they can make you harm yourself.

Rattlesnake On Our Driveway

The rattlesnake is a beautiful snake that demands great respect, but if left alone, it will cause no harm. Leaving them alone becomes a problem if you don't know that they are there and suddenly discover their presence. Not knowing causes one to be extremely cautious during the rattlesnake time of year. Snakes hibernate in the winter so, in rattlesnake infested areas, we reserve certain activities until winter time. Neither Barbara nor I have been bitten by a rattler but both of us have been too close on several occasions. We have had horses, cats and dogs bitten by rattlers. Sometimes dogs just show no respect and rattlers demand respect. Cats are usually just in the wrong place. Normally a horse is aware of the presence of rattlers and intelligent enough to be respectful. One of our horses got bitten when she inadvertently grazed too close to a bush where the snake was waiting for edible prey. I'm sure the horse would have gladly apologized for the interference and moved away before being bitten but rattlesnakes don't accept apologies.

As for me, I know that I'll not die from rattlesnake venom because seeing the snake strike at me will give me a fatal heart attack.

We have lots of birds on the ranch ranging from song birds to buzzards and all have a place. The song birds are here to make our world more pleasant and the buzzards are here to help keep it clean. Cranes habituate our stock ponds and quail scamper through our underbrush. Doves are Barbara's favorite and they calm the countryside with their soothing calls. I'm sure my guardian angel is a dove because her call gets me through many unnerving problems. One winter we had a pair of roadrunners that nested near the house and kept our flower beds clean of small snakes, bugs and worms. We got rather attached to their presence and friendliness. It was fun to watch them come in the yard, catch a small snake, then sit on the fence with the snake in their mouth to let us witness their skill. Barbara was especially distraught one day to find a roadrunner head laying by the walkway. As best we could determine the decapitation was done by a raccoon because the bird's feathers were scattered along the roof of the house. The raccoon may have gotten both of the pair because we never saw either bird again.

Feral cats of the house cat variety have become a regular member of our wildlife. Town folks bring their unwanted cats to the country and leave them to find their own way. Many become acclimated to the country life and fend for themselves. We partially domesticate some of these cats and keep them at our barn to control the rats and mice. Rats and mice are desired food of the rattlesnakes and by controlling their population around the barn helps keep the rattlers away.

The ranch has its share of white tailed deer. The deer are our favorites and we love to watch them. They are not easily domesticated so we mostly see them around the ranch. But they too have habits worth watching and demonstrate an extreme intelligence, just ask any deer hunter.

While all wild animals have intelligence or they would have disappeared long ago, we have especially noted the intelligence of the cats, deer, raccoons and coyotes that live on or frequent the ranch.

Raccoons

Raccoons are attractive animals but that is a very distracting feature. Coons are the most destructive animals on a ranch. They are everywhere and respect no property. They get in the barns and are not satisfied until they have torn something apart.

Raccoon Baby

I wrote the following poem to stress this destructive nature and what we would like to do about him.

Mean Coon

Something needs to happen soon
Like the demise of one mean coon.
His life's in danger I sadly fear
He's done too much round here.

He got in the orchard for his usual treat
And peaches, apricots and plums did eat.
Not just a few that we could spare

He ate them all while he was there.
He wanted in my shop I'm sad to say
But the screen door was in his way.
His paw prints were easily seen
As he madly ripped the screen.

Can't leave feed in my truck
He'll come by, tis my luck.
All those bags, he'll not be done
Til he's torn them one by one.

On the roofs, each cranny and nook
He'll tear off the shingles just to look.
I think I've figured the plan
He's on the staff of the roofer man.

So I've set the trap and can't wait
To see if he'll take the bait.

He'll be mine, if I get my wish
That he can't resist the tuna fish.

Tis the morrow, I'll take a look.
The trap is sprung, the fish he took.
My day is made, I'm on an upper
Now that coon'll be buzzard supper!

But next night another torn sack
That coon couldn't be back
That's for sure, must be some other
Guess he had at least one brother.

So, back to the trap and another dish
Of that strong smelling tuna fish.
He can't resist it is my hunch
And this one'll be buzzard lunch!

Although very destructive, raccoons are born actors. As the poem reveals, sometimes I use a live trap to try to catch the mean coon. One time, our grand sons were on the ranch for a visit. I had set the live trap and sure enough a coon had taken the bait. The grand sons were out trap shooting and wanting to learn how to shoot and how to hunt. I took them to the barn where the coon was trapped and asked them to shoot the coon in the trap so I could haul him away to feed the buzzards. The boys were anxious to go shoot the coon! We approached the trap and the coon decided to show his acting skills for the boys. He laid back and raised his front legs up around his head with a slight grin as if to say "how could you possibly hate a guy this cute?" It worked, the boys took one look and mildly put their guns to their sides and left the barn.

We feed commercially prepared range cubes to our cows. These are ground feed materials that have been molded into small cubes for dispensing onto the ground rather than having to be fed in troughs. As shown earlier in the Dialogue, we haul the cubes in an enclosed trailer that will dispense the cubes into small piles as one drives along. The cows quickly learn to follow the cube trailer and eat the cubes.

The cube trailer is enclosed except for small openings to allow for the chain device to dispense the cubes from the storage bin. Raccoons have front feet that are very similar to hands and it is amazing how they can reach into small places and get things to eat. They love the protein in the cubes and can remove several pounds in one night by reaching in through the small openings. They are not only destructive, they can be very expensive to have around.

Raccoons are very smart. Barb's dad told us a story about the intelligence of raccoons. He said the raccoons would pick the lock on the barn door, eat all they wanted and then come to the house and beat him playing dominoes.

Cats

Probably the hardest working animals on our ranch are the barn cats. Barns are very attractive to mice and rats. Mice and rats are the favorite meals for rattlesnakes. In order to control the influx of rattlers, we use cats to eliminate the part of the food chain that attracts the snakes. I shouldn't have said eliminate, a better word is control. I don't think that mice and rats will ever make the endangered species list.

America is becoming populated with feral cats. All you need to do to put cats at the barn is to domesticate one female cat. The male cats will find her if she is fertile and soon you will have several cats. The species will vary quite a bit as the various male species come around.

As with most animals, some of the cats will be easily tamed and others will always be wild. But even the wild ones respond to regular feeding and will stick around the barn to help control the mice and rats. Some cats are very intelligent and some are admittedly dumb, but both serve the purpose.

Almost daily, one of us will go to the barn and feed the cats. We have pans that stay under a barn shed that we use for feeding the cats. The cats, of course, become rather attached to those pans and surely recognize the purpose of the feed pans. The cats recognize us as their feed masters and will come partly up the road to meet us and escort us back to the barn feed pans. They gather around waiting for the day's allowance and always seem pleased with whatever left over scraps of food that we bring to them.

We do not expect them to depend totally upon our food scraps for their total diet. I'm sure that they forage for themselves, particularly if we are late bringing down the food. On several occasions, if we have been late bringing in the daily treat, we have found bird feathers in the food pans. Obviously, one cat has foraged a bird and brought it in to share with their friends.

Now one might question the intelligence of that cat. Why would it bring its hard caught meal to the food pan and share it with the other cats? It could be that the act was not one of sharing but merely satisfying the idea that the food, whatever the source, should be served from the food pan. Whatever the reason, the feathers are proof that the bird was served from the food pan.

We had one visiting sire that was obviously of Manx breeding because we got a series of bob-tailed cats. One of these bob-tailed

kittens was very aggressive and was a joy to watch as he managed
to bluff his way around the barn. Even the Pyrenese dogs had to
give way as I try to relate in the following poem.

THAT BOB TAIL CAT

Down at the barn, we have some cats.
If we didn't, it'd be full of rats.
The cats we need, are just one or two.
But nature doesn't permit such a few.
One day there'll be not so many,
Then suddenly there's ten or twenty.
Soon we've more than run out of rats.
We got too many of those darn cats.

We started out with a Persian beauty.
She was furry, a long tailed cutie.
Just a lone cat all cute and furry,
For a replacement we needn't worry.
With female cats it's just fate,
Out of nowhere there's a mate.
Before long the kittens'll skitter,
As there's another new litter.

The pure Persian fur is no more.
Now we've got colors galore.
Some are yellow, some are gray,
Some can easily hide in the hay.
Some are cute and some are not.
Whatever new daddy is our lot.
"Hey Barb, look at that
There's a tiny bob-tail cat."

Bobtail Cat

The male cats from the countryside
Are not the kind one takes much pride.
You never know what you're gonna get
When you look for that new kit.
Some are mild and some are bold.
Some you would like to hold.
But this one'll have none of that
"Not me", says the bob-tail cat.

Also around the barn yard
Two Pyrenese dogs stand guard.
They are big as they need be
To guard things successfully.
But it's no secret that
Dogs do love to chase a cat.
Hey what's this, a real spat
He don't run, that bob-tail cat.

When the dogs get in the mood
They come to the barn for some food.
And food they always can
Find in their food pan.
Most cats know and fear
So when the dogs are near
They quickly scatter and hide
Sometimes fear is safer than pride.

One day, the dogs out of sight,
Shyly cats came to steal a bite.
They carefully watch and wait
Fearing that food might be their fate.
And safely most were gone again
When the dogs came running in.
But there in the pan one just sat
Guess who? That bob-tail cat.

Barely a pound is his tare,
Mostly skin, bone and hair.
But much to the dog's dismay,
He's determined to get his way.
He cares not that he might,
Be dog food in just one bite.
With paw raised, he stands his ground
As the dogs nearer bound.

Quickly, the dogs seemed to know
That they best go pretty slow
For here's one who seems bound
Not to give away any ground.
So loudly bark they hoped to scare
Prancing round with bristled hair.
But much to their dismay
The little cat gave no way.

Soon the dogs gave in defeat
And the little cat began to eat.
"My bluff did work at that",
Smiled that darned bob-tail cat.
Now he scampers here and there
Never a fear, never a care.
In his world, he made a place
By meeting challenge, face to face.

Proving again with no ifs or buts
To gain respect takes lots of guts.
It's not stature that brings pride.
Character comes from deep inside.
For Barb and me as our way we earn
From animals we can surely learn.
And with pride, we're glad that
He's our fearsome bob-tail cat.

Black Cats

Some people believe that black cats bring bad luck while others think that black cats bring good fortune. The Egyptians were probably the first to tame cats and use them to keep the rat and mice populations under control. The Egyptians also worshiped the cats and treated them as gods. Anyone who was found guilty of intentionally killing a cat was usually put to death. I don't know how they felt about black cats.

Barbara's mother was very superstitious about black cats. Barb recalls that her mother would turn the car around and take a different route if a black cat crossed the street in front of her. I'm glad that Barb didn't inherit that superstition because we have two beautiful black cats down at the barn. No telling how many times they have crossed the many paths that we take in and around the barn. Come to think of it, they may have been responsible for the bad luck I had when a calf caught my arm between his back and the squeeze chute the other day. Or maybe those black cats are the reason that we are having trouble making the ranch pay its way.

Bad luck or not I like the cats being down there and I like the job they are doing keeping the rat and mice populations under control. I don't know what I would do about it anyway because if I eliminated them the Egyptians might come and get me. If not the Egyptians, the People for the Ethical Treatment of Animals (PETA) surely would come calling. I've heard that cats have nine lives but it would be my bad luck that PETA was keeping count and would claim that I used up their ninth.

Blind Cat

One might expect with many different feline genetics roaming the countryside that there would occasionally be a genetic problem. We had one line of bob-tailed cats that would go blind after they became adults. They would seem to be all right at birth but as they got older their eyesight would fail. It was good that the effect manifested after they had time to learn their surroundings.

We first noticed the trait with one of the young females that was one of our favorites. We began to notice that she would bump into things that should have been familiar to her. It finally dawned on us that she had become totally blind.

Although blind, she would still go out to hunt just as she did in her younger and better-seeing days. We would notice her coming back to the barn, always in the same path. She had learned to sense her way by things that were familiar. She apparently knew that her meow would echo from the wall of the sheet metal barn and she could tell her location from the familiarity of that sound. She would approach the gate to the barn and then reach forward and feel for the familiar opening between the gate boards. Sometimes she would misjudge and miss the gate altogether. She would calmly advance and try again until she found the gate board. Once she located the gate she would maneuver through and advance along the familiar path to the feed pan or other familiar areas under the barn. It was an interesting experience to quietly watch her find her way around without that wonderful sense of sight.

This particular blind cat got around in more ways than one. She got pregnant and finally had her kittens but we never did find them. We don't know whether she could not find them herself to raise them through the dependent stage or whether some varmint got them shortly after birth.

We love having our grand kids come to visit. Animals are interesting to grand kids, particularly those who are growing up in the cities without the day to day experience of ranch life. We have four grand kids, two granddaughters and two grandsons. The blind cat and her experience with birthing and losing her kittens was particularly interesting to Donna when she was about 10 years old. Donna noticed that the other cats would help the blind cat by kindly showing her around. After one summer visit, Donna went home and wrote this poem.

DARKNESS, THAT LITTLE BLIND CAT
By Donna Young

There is a cat who can't catch those darn rats.
She walks around bumping and hitting.
When the food comes out we say,
"look! there she is sitting, sitting.
Even though she is sweet and kind
Not like most cats, she is blind.

That little gray cat helps her all the day,
Up in the barn and around the hay.
We think she has some little kitties
But we can't find them, what a pity.
She'll be a fine mother, we have no doubt
We will find them when they shout.

Those other cats treat her fine no doubt about that
Mostly she is the same as the others, she is a cat.
And she could see before
But not any more.

She is scared that creatures might hurt her kitties
She is blind and they are so itty bitty.
She might not have had her kitties
If she hasn't that would be a pity,
But if she has had her cats
She would stand proudly, not in defeat.
She will protect them in the cold or in the heat.
The other cats think her odd, but sweet and kind,
She is our cat even though she is blind.

There were several cats in this blood line that would become blind as they reached adulthood. The blood line was especially vulnerable and we no longer notice any of that sad genetic experience.

One of the joys of having cats around the barn are the cute little kittens and with several momma cats, we have a bunch of those kittens. Like the mommas, some are cute and some are not but all are a joy to watch.

Actually they are quite vulnerable to the woes of cathood. Nature seems to know how to control overpopulation. I guess they are easy victims of the world of predators that roam the countryside. For whatever reason, our cat population stays fairly uniform and we usually have a sufficient number to keep the mice and rats under control.

Most of the kittens are pretty leery of humans but some will let you get close to them after they have spent a few days around us while we are doing our barn chores. The feed-pan is also an attraction that tends to tame the wild. I found one little kitten hunkered down in a box that I had built around a water faucet to keep it from freezing. It seemed so harmless, not too frightened, so I decided I would pick it up and gentle it a bit.

This poem gives an account of what happened.

KITTEN BITTEN

Real cute, that little cat
Who partly hidden sat.

I sure would like to pet'em.
So in a wink
Little did I think,
As I reached down to get'em.
I got'em you see
Or rather—he got me.
Now I've been kitten bitten!

With all those claws
On all four paws
And teeth as sharp as a tack.
He twisted and turned
A lesson I learned,
A lesson he taught, in fact.
Not long did it take
My mind to make.
I think I'll put him back!

Deer

The prettiest animals on the ranch are the white tailed deer. They spend most of their time out of sight. They see us daily and are accustomed to us and to our vehicles that we use on the ranch. Barbara and I see them daily because they consider us to be part of nature.

Deer Out Back

It seems sort of odd but even though the deer live here on our ranch they do not belong to us, they belong to the State of Texas. We are just honored that the State's deer chose our ranch as a place to live, forage and reproduce. We provide the deer forage, water, feed, mineral and a place to hide but we do not get any pay from the State.

The State encourages us to prevent overpopulation by allowing deer hunting, provided that we pay a fee to the Texas Parks and Wildlife Department for that privilege. All we get in return for that fee is the services of a Game Warden who patrols for illegal hunting and ensures the State that we have paid our fee. If we want to take a deer for our own use, we must first get a hunting license by paying an additional fee. The only pay we get is the enjoyment

Deer In Yard

of observing these beautiful creatures in their natural habitat and that is payment enough.

Our house is in one of the larger pastures on the ranch and the deer can come to the house at will. To encourage them to come so that we can see them and watch another species of dumb animals display their intelligence, we began putting shelled corn out behind our house for them to eat. We put the corn out about an hour before dark so that we can have some peaceful time to watch their activities. This practice has been a joy indeed.

It didn't take long for the deer to discover the feed and get accustomed to coming near the house. We stay in the house and watch from my office windows. They see us inside but have learned that we will not harm them from that viewpoint. If times have been bad and there is a lack of feed in the pastures, they may come for their corn a little early. If we don't get the feed out when they think it should be, they will circle the house and look in at us as if to say "yoohoo, here we are".

We now have about twenty doe that come each evening. They have their pecking order and each deer in the group knows who is in charge. The boss doe will approach another, back her ears and wait for the other doe to give way. If the other doe doesn't yield, the boss will rise to her hind feet, take the stance of a boxer, and hit the other doe with her front feet. Believe me, it's not a gentle nudge, she means business and usually gets response. Occasionally, if the lesser doe does not respond, there will literally be a boxing match until one yields.

We can tell when the does have given birth and we begin to watch for them to bring in their fawn. It takes awhile for some of the doe to get brave enough to risk bringing in the fawn when they come to eat. Some are more trusting and will first bring the fawn near and bed it down while they come in to eat. Before long, the fawns gain courage and will run ahead of the mommies to the feed. As the fawn grow, it becomes obvious which ones are the males because they will begin to grow small, velvet covered knobs that will eventually be their antlers. The yearling does will continue to come in to eat with the mothers and so will the yearling males until they form significant antler. Once they begin to show male deer traits, they are no longer welcome with the females and are forced to go out on their own. As with most animal species, the stag deer that have sired the fawns have no part to play in the upbringing of the offspring.

Wild animals are on constant alert in order to protect themselves from the various dangers. Deer are very alert and if one sees or hears something suspicious she will stomp her foot and snort. Upon that warning, the group doesn't wait to see the danger, they immediately scatter and only when they feel safe will they search for the danger. If they respond to a false alarm, they will return with a cautious, stiff-legged gait and continue to eat. The survival instinct is to respond first and then investigate.

A Dear Deer

One day I was out on the four wheeler and came upon a deer hung in the fence. It was a young buck that had caught his right hind leg in the top wire as he tried to jump the fence. His foot was about five feet off the ground. He was very scared of me when I approached and my calming talk didn't seem to ease his fear. It took me a while to get the wire loose enough to remove his foot so by the time I got his foot free he had calmed considerably. Once clear of the wire, I set him free and stood back to watch his recovery.

I had continued my calming talk during the ordeal and I was amazed at what I then experienced. He had turned to look at me as he stood just a short distance away. He stood still and continued to stare as I continued to talk. His dark eyes were so beautiful as he stared at me. We both stood still for several minutes and continued our conversation. I told him that I only wanted to help and that Jesus loves him. I sensed that he was trying to tell me that he appreciated my help and was extending his love.

Experiences such as this are precious moments in the country and always makes me thankful that God created this wonderful world and all its occupants.

Dumb animals are not so dumb.

Coyotes

No doubt the smartest animals on the ranch are the coyotes. Well maybe they compete with the horses for that intelligence status.

Coyotes are so smart that they have managed to get the government to protect them even though they are devastating the sheep and goat industries. We are not sure how they managed this feat but it no doubt proves that they are smarter than the bureaucrats. When bureaucrats protect wild animals to the extent that they destroy the food and fiber supplies for humans, then it is very clear which of the animals are the least smart.

Coyotes are a natural predator of sheep and goats. They are nocturnal and very wily. It is difficult to trap them because they quickly learn how to avoid the danger of traps. They seldom live in the area that they are predating. Usually they travel several miles to make their kills at night and then return home before daylight. They spend the day resting up for the next night much to the ire of the rancher who suffered the kills of the night before.

Coyotes are not conservative in their kills. They will kill and eat the heart and liver and leave the rest of that lamb or kid goat and go for another kill. It is especially bad when they are teaching their young how to hunt. They may make several kills as they perform their lesson plan for that night. They will make a kill and leave that fresh carcass to take up the chase as another lesson for their pups. Each hunt is for a fresh carcass as they never go back to eat the remains of a previous nights work

Coyotes have few natural enemies so there are few controls on their population growth. The primary controls are lack of food and disease. They are prone to contract rabies and become a direct threat to humans, particularly around cities and towns. As their rural habitat becomes devoid of attractive food they move closer into towns and cities. They will attack dogs and cats in the cities and spread rabies to the family pets. They are a definite threat to the health of humans.

Ranchers do not attack wild animals just to eliminate wild animals. Ranchers do try to control animals that are predators of the livestock that are being raised as produce of the ranch. If the coyote population is small and not a serious threat to the ranch operation then they will be left to natures ways. When the population increases and the number of kills become a problem then the rancher will provide the essential control. If allowed to control predator populations, the rancher can stabilize the populations and thereby lessen the health threat to nearby cities and towns. When government regulations remove the logical means of predator control, the system fails and not only is the food and fiber production put in jeopardy but so is the health of humans.

Turkey

Turkey

Wild turkey are a beautiful addition to the our ranching business. Its not easy to tell whether they are intelligent or dumb but I choose intelligent.

It is interesting that in some animal species the male is more beautiful than the female. The tom turkeys are magnificent. They spread their feathers and strut around for all the hen turkeys to see their beauty. What ever the reason it seems to work. The hens eventually lay their eggs and a new flock of turkey are brought into being.

So far as I know, turkeys are not a harm to the ranch. They do not deprive the other animals of food or cause any concern to the neighboring animals. They accept their place and we see them amongst the deer and cattle as they make their own way.

Leasing the ranch for hunting helps provide more funds to keep the ranch operating. Turkeys and deer are the primary reason the hunters pay their fees and come to visit. Sometimes the hunters are just interested in getting out of the big cities and come to enjoy the country beauty.

I am sure glad that when God created human beings he gave the beauty to the females. While some men seem to enjoy strutting around like tom turkeys, I have no desire to do so and am always happy to see the beauty of our women.

CONCLUSION

Barbara has a bumper sticker on the back of our pickup that says

> And to care for and protect his creation
> **GOD MADE RANCHERS**

I hope that sums up the purpose of this book.

Animals, including humans, have a place in this wonderful world that God has created. Living amongst the animals, ranchers respect their intelligence and stay in the ranching business, not just to make a profit, but because they respect and care for this wonderful world and the right of animals of all kinds to be on this earth.

As I stated in the beginning of this book, we hear a lot about animal rights and how cruel ranchers are endangering animal species. We hope that we have shown that ranchers are most respectful of the animals that provide so much enjoyment on this earth.

Our "dumb" animals are intelligent.

ABOUT THE AUTHOR

Kenneth W. Bull was born December 4, 1930 in Brown County, Texas as the tenth of eleven children. His parents owned a small farm and typically raised their own labor force. His agriculture experience started early.

While attending Arlington State University in 1951, the draft exemption for college students was removed and Ken decided to enlist in the Air Force rather than be drafted into the Army. He was trained as a control tower operator and immediately applied for flight training. In 1952, he received his commission as an officer and his wings as a Radar Observer.

Ken and Barbara Bell were married June 28, 1952 and began their joint venture in the Air Force at Elmendorf Air Force Base, Anchorage, Alaska. Over the next twenty years, they experienced many assignments. The Air Force provided Ken with a Bachelor Degree in Engineering and a Masters Degree in Physics. Most of his career involved weapons research. Ken retired from the Air Force in 1971 as a Lt Colonel.

Ken and Barbara have been managing Barbara's ranch since 1974 and have enjoyed these many years in the wonderful world of animals.